Name: Eng. / Mostafa Yacoub Abdellatif Mahmoud

Nationality: Egyptian

ORCID: 0000-0002-9991-4624

Email:

moshhaabma2015@gmail.com

Qualification: civil engineer Cairo University 2003

- **<u>Prime and composite numbers:</u>**

In this paper or research, we will discover the relation between prime numbers and the array of odd numbers

$$\begin{vmatrix} 1 & 3 & 7 & 9 \\ 11 & 13 & 17 & 19 \\ 21 & 23 & 27 & 29 \\ 31 & 33 & 37 & 39 \\ 41 & 43 & 47 & 49 \\ 51 & 53 & 57 & 59 \end{vmatrix}$$

And so on....

based on my discovered formula that connects prime and composite numbers.

- ## My discovered formula:
Definitions:
Array PTBP

It is the following Array of odd numbers

$$\begin{vmatrix} 1 & 3 & 7 & 9 \\ 11 & 13 & 17 & 19 \\ 21 & 23 & 27 & 29 \\ 31 & 33 & 37 & 39 \\ 41 & 43 & 47 & 49 \\ 51 & 53 & 57 & 59 \end{vmatrix}$$

And so on....

- For a given set of consecutive primes whose numbers =n that start with prime 3 and end with prime F and not including prime 2 and prime 5
i.e.

set=[3,7,11,13,……………………………………
…………………….,F]

S=product of those consecutive primes

i.e

$$S = \prod_{i=3}^{i=F} (i)$$

Range=R_k = 10 × S × k

Where k = [1, 2, 3, 4, ...,
∞(infinity)

i.e R_1=10 x S x 1 and R_2=10 x S x 2

And so on

- **Number of composite numbers that belong to Array PTBP and created by the effect of those consecutive primes within the range R_K**

- =[(K × $4^{\times \frac{S}{3}}$) + (

$$\sum_{j=7}^{j=F} (K \times 4 \times (\frac{S}{j}) \times$$

$i = prime\ number\ befor\ current\ prime\ number\ j$

$$\prod_{i=7} \qquad (\frac{i-1}{i})$$

)]-(n)

Where j =consecutive values of primes

7, 11, 13,..............., F

And i= consecutive values of primes

3, 7, 11, 13,........., prime before current j prime

The previous formula can be applied for any number of consecutive prime numbers that start with prime number 3

- The first term $(k \times 4 \times \frac{S}{3})$ represents the count of unique Composite numbers +1 that belong to the Array PTBP and are created by prime number 3 within the range

$$R_k = 10 \times S \times k$$

- **The second term**

$$\sum_{j=7}^{j=F} \left(K \times 4 \times \left(\frac{S}{j}\right) \times \right.$$

$i = prime\ number\ before\ current\ prime\ number\ j$

$$\prod_{i=7} \qquad \left(\frac{i-1}{i}\right)$$

Represent the count of unique Composite numbers+n-1 that belong to the Array PTBP and are created by each prime number after the prime number 3 within the range
$R_k = 10 \times S \times k$

- **The third term (-n)**
Subtracting n (number of consecutive primes starting from prime number 3) because the count of composite numbers generated from those consecutive primes includes the count of those primes in the range

$$R_k = 10 \times S \times k$$

- Explanation and proof for my theory in my previous paper (prime number theory)
- We will mention only the concept of number cycle

We can use the number cycle concept to understand the behavior of consecutive primes in creating composite numbers.

i.e.

$$S = \prod_{i=3}^{i=F} (i)$$

Range=cycle range= $R_k = 10 \times S \times k$

Where k= [1, 2, 3, 4, ..., ∞(infinity)

i.e. $R_1 = 10 \times S \times 1$ and $R_2 = 10 \times S \times 2$

And so on

- Now consider only one k value =1
And Now
- For any set of consecutive primes

i.e.

set=[3,7,11,13,...........................

.........................,F]

S=product of those consecutive primes

i.e

$$S = \prod_{i=3}^{i=F} (i)$$

Range= Rk = 10 × S×k

- Any cycle containing two types of numbers
- First type (numbers that are divisible by prime numbers within the set).
- The second type (numbers that are not divisible by prime numbers within the set).

- For example, the following figure (considering the set of consecutive primes=[3,7]) shows the two types of numbers the colored one represents the first type while the uncolored numbers represent the second type

- We can see there must be a repeated pattern for each type of number and for both of them together each cycle up to infinity i.e for

k= [1, 2, 3, 4, ……………, ∞(infinity)

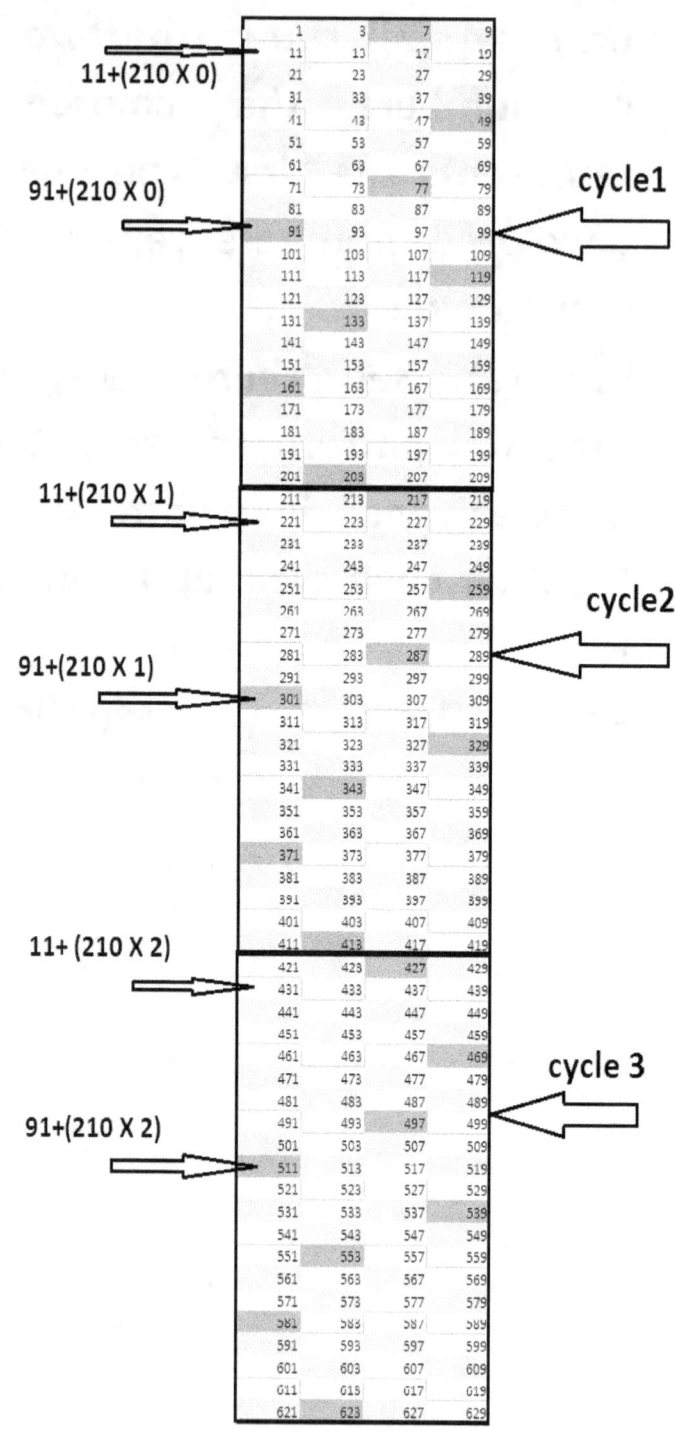

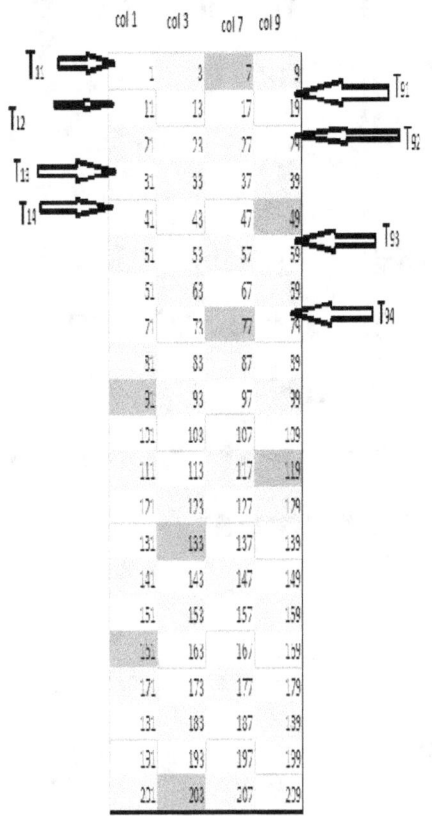

In the previous figure, we called each column

col 1, col 3, col 7, and col 9 concerning the last digit of each column respectively to arrange the uncolored values in the first cycle

T_{1i} , T_{3i}, T_{7i}, T_{9i} where i takes values within range

$$=1: \prod_{j=3}^{j=F} (j-1)$$

j takes value of prime numbers that belong to set of consecutive primes 3,7,11,....F

Let p= next prime number after prime number F

$$P \times \left(\sum_{i=1}^{i=\prod_{j=3}^{j=F}(j-1)} (T_{1i} + T_{3i} + T_{7i} + T_{9i}) \right)$$

= P x

$$\left\{ \left(\left(1 \, x^{\prod_{i=3}^{i=F}(i)} \right) + \left(10 \times \sum_{j=1}^{j=\left(\prod_{i=3}^{i=F}(i)\right)-1} (j) \right) \right) \right.$$

$$+ \left(\left(3 \, x^{\prod_{i=3}^{i=F}(i)} \right) + \left(10 \times \sum_{j=1}^{j=\left(\prod_{i=3}^{i=F}(i)\right)-1} (j) \right) \right)$$

$$+ \left(\left(7 \, x^{\prod_{i=3}^{i=F}(i)} \right) + \left(10 \times \sum_{j=1}^{j=\left(\prod_{i=3}^{i=F}(i)\right)-1} (j) \right) \right)$$

$$+ \left(\left(9 \times \prod_{i=3}^{i=F} (i)\right) + \left(10 \times \sum_{j=1}^{j=\left(\prod_{i=3}^{i=F} (i)\right)-1} (j)\right)\right)\}$$

$$\times \left(\prod_{i=3}^{i=F} \left(\frac{i-1}{i}\right)\right)$$

$$= P \times \left(\left(20 \times \prod_{i=3}^{i=F} (i)\right) + \left(40 \times \sum_{j=1}^{j=\left(\prod_{i=3}^{i=F} (i)\right)-1} (j)\right)\right)$$

$$\times \left(\prod_{i=3}^{i=F} \left(\frac{i-1}{i}\right)\right)$$

i.e

$$\left(\sum_{i=1}^{i=\prod_{j=3}^{j=F} (j-1)} (T_{1i} + T_{3i} + T_{7i} + T_{9i})\right)$$

$$= \left(\prod_{i=3}^{i=F} \left(\frac{i-1}{i}\right) \times \left((20 \times \prod_{i=3}^{i=F} (i)) + \left(40 \times \sum_{j=1}^{j=\left(\prod_{i=3}^{i=F} (i)\right)-1} (j)\right)\right)\right)$$

i.e

$$\left(\sum_{i=1}^{i=\prod_{j=3}^{j=F}(j-1)} {}^{(T}{}_{1i} + T_{3i} + T_{7i} + T_{9i}) \right)$$

$$= \left(\left(20 \text{ x } \prod_{i=3}^{i=F}{}^{(i-1)}\right) + \left(40 \text{ x } \sum_{j=1}^{j=\left(\prod_{i=3}^{i=F}(i)\right)-1} {}^{(j)}\right) \text{ x } \prod_{i=3}^{i=F}{}^{(\frac{i-1}{i})}\right)$$

$$= \left(\left(20 \text{ x } \prod_{i=3}^{i=F}{}^{(i-1}\right) \right) + \left(40 \text{ x } \prod_{i=3}^{i=F}{}^{(\frac{i-1}{i})} \text{ x } (1/2) \text{ x } \left(\left(\prod_{i=3}^{i=F}(i)\right)-1\right) \text{ x } \prod_{i=3}^{i=F}{}^{(i)}\right)$$

$$= \left(\left(20 \text{ x} \prod_{i=3}^{i=F}{}^{(i-1}\right) \right) + \left(20 \text{ x} \prod_{i=3}^{i=F}{}^{(i-1)} \text{ x } \left(\left(\prod_{i=3}^{i=F}(i)\right)-1\right)\right)\right)$$

$$= \left(\left(20 \, x \prod_{i=3}^{i=F} {}^{(i-1)} \right) \right) - \left(20 \, x \prod_{i=3}^{i=F} {}^{(i-1)} \right) \right)$$
$$+ \left(20 \, x \right.$$

$$\left. \left(\prod_{i=3}^{i=F} {}^{(i-1)} \right) x \left(\prod_{i=3}^{i=F} {}^{(i)} \right) \right)$$

- i.e

$$\left(\sum_{i=1}^{i = \prod_{j=3}^{j=F}(j-1)} (T_{1i} + T_{3i} + T_{7i} + t_{9i}) \right)$$

$$= 20 \, x \left(\prod_{i=3}^{i=F} {}^{(i \, x \, (i-1))} \right) = \text{summation of}$$
uncolored numbers

- So we can get the summation of colored numbers
Which

$$= \left(\left(20 \; x\prod_{i=3}^{i=F}(i)\right) + \left(40 \; x \sum_{j=1}^{\left(\prod_{i=3}^{i=F}(i)\right)-1}(j)\right)\right)$$

$$20 \; x\left(\prod_{i=3}^{i=F}(i \; x \; (i-1))\right))$$

$$= \left(\left(20 \; x\prod_{i=3}^{i=F}(i)\right) + \left(40 \; x \; (1/2) \; x\right.\right.$$

$$\left(\left(\prod_{i=3}^{i=F}(i)\right)-1\right) \quad x \quad \prod_{i=3}^{i=F}(i)\right) \quad - \quad 20 \; x$$

$$\left(\prod_{i=3}^{i=F}(i \; x \; (i-1))\right)))$$

$$= \left(\left(20 \; x\prod_{i=3}^{i=F}(i)\right) + \left(20 \; x\prod_{i=3}^{i=F}(i) \; x\prod_{i=3}^{i=F}(i)\right)\right.$$

$$- \left(20 \quad x\prod_{i=3}^{i=F}(i)\right) - 20 \; x\left(\prod_{i=3}^{i=F}(i \; x \; (i-1))\right))$$

$$= 20 \; x \left(\left(\prod_{i=3}^{i=F}(i x i)\right) - \left(\prod_{i=3}^{i=F}(i \; x \; (i-1))\right))\right)$$

- **So the summation of values of the cycle**

$$= 20 \; x \; \left(\prod_{i=3}^{i=F}(i x i)\right)$$

- **For example**

1	3	7	9
11	13	17	19
21	23	27	29
31	33	37	39
41	43	47	49
51	53	57	59
61	63	67	69
71	73	77	79
81	83	87	89
91	93	97	99
101	103	107	109
111	113	117	119
121	123	127	129
131	133	137	139
141	143	147	149
151	153	157	159
161	163	167	169
171	173	177	179
181	183	187	189
191	193	197	199
201	203	207	209

- **The figure represents one cycle for the consecutive primes 3,7**
  The summation of all numbers within the figure
  $20_x (3*3*7*7) = 8820$

- And the summation of uncolored numbers which is not divisible by 3 & 7 =20 x ((9-3)*(49-7)) = 5040

- The summation of colored numbers which are divisible by 3 or 7 = (20*((9*49)-((9-3)*(49-7))) = 3780

- This proved formulas can be used for the first cycle for any set of consecutive primes
  And those formulas are great explanations for the connection of prime numbers and the array of odd numbers PTBP

$$\begin{vmatrix} 1 & 3 & 7 & 9 \\ 11 & 13 & 17 & 19 \\ 21 & 23 & 27 & 29 \\ 31 & 33 & 37 & 39 \\ 41 & 43 & 47 & 49 \\ 51 & 53 & 57 & 59 \end{vmatrix}$$

And so on....

- **Now we will make a summary of all equations based on the figure below**

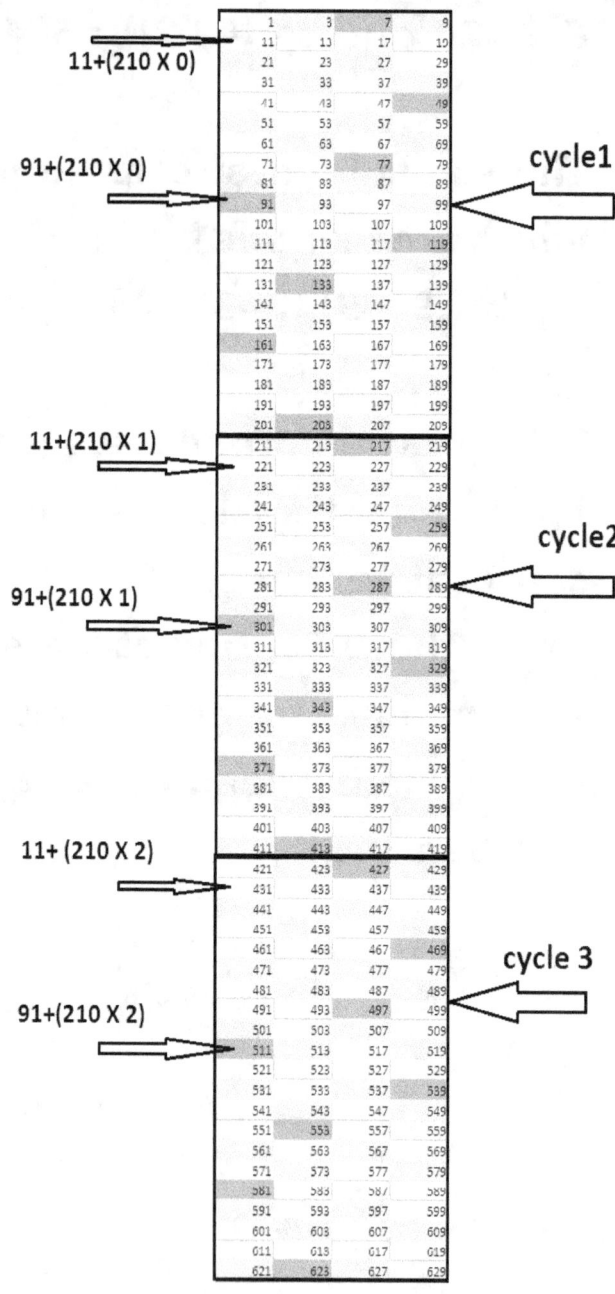

- Previous figure represent three cycles (K= 3) for the set of consecutive primes set=[3,7]
- The yellow cells represent the numbers that are divisible by prime 3
- The green cells represent the numbers that are divisible by the prime number 7
- As we note the number 21 is colored yellow because prime number 3 has priority over prime number 7

- The count of composite numbers produced by prime number 3 within the first cycle

$$= \left[k \times 4 \times \frac{S}{3} \right] - 1$$

$$= \left[1 \times 4 \times \frac{3 \times 7}{3} \right] - 1 \quad =27$$

- we subtracted 1 because the count $\left[k \times 4 \times \dfrac{S}{3} \right]$ include prime number 3 itself only in the first cycle

- The count of composite numbers produced by prime number 3 within the only second cycle

$$= \left[k \times 4 \times \dfrac{S}{3} \right] =$$

1

$$\times 4 \quad \times \dfrac{3 \times 7}{3} \quad = 28$$

- The count of composite numbers produced by prime number 3 within the first and second cycle

$$= \left[k \times 4 \times \dfrac{S}{3} \right] - 1$$

$$=\left[2 \times 4 \times \frac{3 \times 7}{3}\right]-1 \quad =55$$

- The count of composite numbers produced by prime number 3 within the first second and third cycle

$$=\left[k \times 4 \times \frac{S}{3}\right]-1$$

$$=\left[3 \times 4 \times \frac{3 \times 7}{3}\right]-1 \quad =83$$

- The count of composite numbers produced by prime number 7 within the first cycle

$$=\left[\left(k \times 4 \times \left(\frac{S}{j}\right) \times \prod_{i=3}^{i=3}\left(\frac{i-1}{i}\right)\right)\right]-1$$

$$= [1 \times 4 \times \frac{3 \times 7}{7} \times \left(\frac{3-1}{3}\right)] - 1$$

$$= 8 - 1 = 7$$

- The count of composite numbers produced by prime number 3 within the only second cycle

$$= [1 \times 4 \times \frac{3 \times 7}{7} \times \left(\frac{3-1}{3}\right)] = 8$$

- The count of composite numbers produced by prime number 7 within the first and second cycle

$$= [2 \times 4 \times \frac{3 \times 7}{7} \times \left(\frac{3-1}{3}\right)] - 1 = 15$$

- The count of composite numbers produced by prime number 7

within the first second and third cycle

$$=[3 \times 4 \times \frac{3 \times 7}{7}$$

$$\times \left(\frac{3-1}{3}\right)] - 1 \quad =23$$

- If the set of consecutive primes =[3,7,11]
  Then
- The count of composite numbers produced by prime number 11 within the first cycle

$$= \left[\left(k \times 4 \times \left(\frac{S}{j}\right) \times \prod_{i=3}^{i=7} \left(\frac{i-1}{i}\right)\right)\right] - 1$$

$$= [1 \times 4 \times \frac{3 \times 7 \times 11}{11}$$

$$\times \left(\frac{3-1}{3}\right) \times \left(\frac{7-1}{7}\right)] - 1$$

= 48 - 1 = 47

- The count of composite numbers produced by prime number 11 within the only second cycle

$$= [1 \times 4 \times \frac{3 \times 7 \times 11}{11} \times \left(\frac{3-1}{3}\right) \times \left(\frac{7-1}{7}\right)] = 48$$

- The count of composite numbers produced by prime number 11 within the first and second cycle

$$= [2 \times 4 \times \frac{3 \times 7 \times 11}{11} \times \left(\frac{3-1}{3}\right) \times \left(\frac{7-1}{7}\right)] - 1 =$$

96 - 1 = 95

- The count of composite numbers produced by prime number 11 within the first second and third cycle

$$= [3 \times 4 \times \frac{3 \times 7 \times 11}{11} \times \left(\frac{3-1}{3}\right) \times \left(\frac{7-1}{7}\right)] - 1 =$$

144 - 1 = 143

www.ingramcontent.com/pod-product-compliance
Lightning Source LLC
Chambersburg PA
CBHW071218290526
45796CB00008B/289